Skilled Helping Around the World
Addressing Diversity and Multiculturalism

Gerard Egan
Loyola University of Chicago

with assistance from
Richard McGourty and Hany Shamshoum

THOMSON
BROOKS/COLE

Australia • Brazil • Canada • Mexico • Singapore • Spain • United Kingdom • United States

© 2006 Thomson Brooks/Cole, a part of The Thomson Corporation. Thomson, the Star logo, and Brooks/Cole are trademarks used herein under license.

ALL RIGHTS RESERVED. No part of this work covered by the copyright hereon may be reproduced or used in any form or by any means—graphic, electronic, or mechanical, including photocopying, recording, taping, Web distribution, information storage and retrieval systems, or in any other manner— except as may be permitted by the license terms herein.

Printed in Canada

1 2 3 4 5 6 7 09 08 07 06 05

Printer: Webcom

ISBN 0-495-09229-0

For more information about our products, contact us at:
Thomson Learning Academic Resource Center
1-800-423-0563

For permission to use material from this text, contact us by:
Phone: 1-800-730-2214
Fax: 1-800-730-2215
Web: http://www.thomsonrights.com

Thomson Higher Education
10 Davis Drive
Belmont, CA 94002-3098
USA

✦

Gerard Egan is the author of The Skilled Helper.

Dr. Richard McGourty is a clinical psychologist and organization consultant.

Mr. Hany Shamshoum, a graduate of Notre Dame University, is from Nazareth.

Table of Contents

PREFACE

This booklet is about diversity with an emphasis on cultural diversity. More importantly, it is about the ways such diversity is found in and gives color to the lives of individual counseling clients. Finally, it is about the kinds of skills, competencies, and values helpers need to understand, appreciate, and work with the diversity that is found in every human being.

A DESCRIPTION OF CULTURE

A group's culture refers to "the way we do things here." The shared beliefs and assumptions of a group interact with the group's shared values to produce group norms. These beliefs, values, and norms constitute the "thinking" part of culture. But culture is ultimately a "doing" thing. Culture is also a set of norms, translated into rules, guidelines, imperatives, habits, regulations, customs, rituals, and the like. Finally, these norms drive patterns of group behavior. The "bottom line" or "doing" part of culture is comprised of the group's patterns of internal and external behavior— both internal; that is, the way members of the group tend to think, see the world, plan, imagine, dream, and so forth, and external; that is, the way they tend to act in public.

Culture, as described here, is not limited to ethnic culture, though most of the examples in this brochure are related to ethnic culture. Governments, businesses, organizations, institutions, associations, professions, and other groupings all have cultures —their assumptions, beliefs, values, and norms all drive patterns of behavior —"the way we do things here." And culture is for better or for worse. Let's start with governments. In the United States it takes about five days to get permission to start a new business. In some countries bureaucratic agencies put up so many obstacles that it takes more than two years. And economic growth suffers (Moffett & Samor, 2005). In universities the cultures or both teaching and research are often at odds with the culture of administration. And the cultures of the helping professions differ from those of physics and engineering. The cultures of Exxon and IBM are very different, but so are the cultures of Oracle and Microsoft even though they are in the same industry. Individuals often find themselves at odds with the cultures or the organizations, institutions, and communities in which they live and work. I have counseled many individuals whose makeup and values put them at odds with organizations where they worked. Accommodation was not easy. So culture is a very rich and useful term. Cultures are both enhancing and limiting, life-giving and infuriating.

DIVERSITY

In what ways do individuals differ? In many different ways, including abilities, age, career, class, color, community, country of origin, country of residence, culture, developmental stage, disability, education, ethnicity, family structure, gender, health status, life history, intelligence, marital status, mental characteristics, national origin, personality variables, physical characteristics, political context, preferences, race, religion, sex, sexual orientation, socioeconomic status, spiritual values, temperament, vocation, and work. And this does not exhaust the pool of diversity factors. Indeed, there is no need to romanticize the *uniqueness* of individuals. Given the combinations and permutations among the variables just named, uniqueness is inevitable.

In human affairs, diversity if often a source of friction. In an interesting experiment, researchers had people come into a room to fill out some forms. As they entered the room the researchers gave them, randomly, a name tag with a blue border together with a blue pen or a name tag with a red border together with a red pen. The participants did not know one another and they did not interact while filling out the forms. Later, the researchers interviewed the participants individually about the forms and how they felt about filling them out. They were also invited to comment on the other participants. A theme emerged: The "blues" felt uncomfortable or did not particularly "like" the "reds." And vice versa for the reds. An artificial bit of meaningless diversity sowed some seeds of discord. History continually shows us how more significant kinds of diversity can lead to horrific bloodshed.

On the other hand, human diversity can be a source of richness. There is hardly a company of any size these days that does not have some form of diversity program. While workplace fairness is an issue here, at the heart of these programs is the conviction that diversity that is duly appreciated, tapped into, and promoted leads to both a better workplace and better business outcomes. Cynics say that this is a lot of nonsense. For them, diversity programs are thinly disguised affirmative action programs. At the other end of the continuum, some pie-in-the-sky idealists give the impression that diversity itself will lead to better results. In my experience, workplace fairness with a tinge of affirmative action and workplace productivity are both factors, no matter what the rhetoric is. Beyond the workplace diversity in itself is, for many people, a celebration of the richness of our humanity.

Commonalities among people are also important. The most important, of course, is our basic humanity. If, instead of differences, this cardinal commonality were the starting point of human interaction, the world would be a different place. All sorts of social institutions are based on commonalities—clubs, neighborhoods, associations. People like doing things with like-minded people.

DEVELOPING A SENSE OF THE WORLD

Over the years my work has taken me to many different countries and I have been privileged to encounter many different cultures. It is, of course, an understatement to say that I learned a lot. Rather these countries together with their cultures and subcultures have contributed enormously to my education and development as a human being and a professional. Specifically, my travels and encounters have helped me develop *a sense of the world*. For me, and, I believe, for everyone, knowing what the world is really like in terms of both its wonders and its horrors is extremely important. Of course you don't have to do the kind of traveling I have to develop a sense of the world. People, books, selective media, and various organizations are the route most people take. In this booklet I use snapshots of my experiences around the world not to treat you to a travelogue but to bring to life lessons about diversity and multiculturalism. And you don't have to travel outside the country to encounter multicultural diversity and all the other forms of diversity. It is all around us.

Picture the following scenario. Students throughout the world in primary, secondary, and tertiary educational settings all learn what the world is really like. They learn about different countries and cultures, including cultures within their own countries; they learn the basics of world economics; they learn about what brings peoples together, what keeps them apart, and what keeps them at one another's throats. Exposed to this kind of multicultural richness, they learn both to appreciate their own culture more fully and to understand and challenge its shortcomings. They prize and learn from differences rather than look down upon or "tolerate" them. In sum, they develop a sense of empathy for the world and their place in it. Without abdicating their own deepest beliefs and values, they learn that there are many ways of living life and come to realize that one way is not necessarily "better" than another. They learn that both good and evil abound in the world and that rooting out evil should start with themselves and their society. Finally, what they learn in school is supported at home, at places of worship, at work, and in other social settings. Such a scenario would, I believe, be good for the students, good for the countries in which they live, and good for the world. This is not a formula for solving the world's problems, but it is a starting point.

You have to work at developing a balanced sense of the world. In the United States both "liberals" and "conservatives" want you to adopt their sense of the world, each camp stridently overstating its case. Since many TV programs mix news, ideology, and entertainment, producing "infotainment" that is anything but balanced, it is difficult getting a sense of one's own country, much less of the world. If you want facts, thoughtful analysis, and honest debate instead of sound bites, Web bytes, and hype in your quest to find out what your own country and the world is really like, you have to search for them. But you don't have to travel the world (physically) to do so.

CULTURE AND DIVERSITY

Culture, one form of diversity, can be an extremely important variable in counseling. Counseling an ethnic Albanian immigrant striving for permanent residency in the United States is not the same as counseling a second-generation Moroccan Arab-American running her own graphic design business. However, as important as culture is, it is just one form of diversity. As the APA guidelines put it, "All individuals exist in social, political, historical, and economic contexts, and psychologists are increasingly called upon to understand the influence of these contexts on individuals' behavior" (p. 377). Culture is one of these contexts. People differ in all sorts of ways. In the end, however, we do not counsel classes or types of people. We counsel individuals. In Japan, where racial homogeneity is the norm, there are enormous differences from subgroup to subgroup and from person to person. I worked with two Japanese men, both tertiary education teachers and both from the same region (far from Tokyo). Yet they were strikingly different. Even though I was the oldest of the three, the younger of the two and I were much more alike, had much more in common, and connected more easily than the two of them. Interests, personality traits, values, and approach to life seemed much more important in our interactions than race and Japanese culture.

Diversity and Multicultural Education in Counseling

The American Counseling Association, the American Psychological Association, the Council on Social Work Education, and the National Organization for Human Service Education have all published guidelines on multicultural education, training, and practice. The spirit of these guidelines is meant to inform what is written in these pages. Though the multicultural movement has taken psychology and counseling by storm — to judge by the massive number of publications in the last few years and, at times, their strident nature—there is not one voice. In the midst of this turmoil, guidelines from these various organizations show evidence of a tempered voice. The cultural guidelines outlined by the helping professions are careful not to overstate the case for culture, minimize its importance, or take sides in the culture wars. Nor are the guidelines presented as a substitute for common sense. For instance, the APA says that its guidelines "are not intended to be mandatory or exhaustive and may not be applicable to every professional and clinical situation. They are not definitive and they are not intended to take precedence over the judgment of psychologists" (*American Psychological Association*, May 2003, p. 378). Finally, the guidelines these associations offer apply to all kinds of diversity. At times the multicultural movement is intelligent and driven by a desire to provide appropriately designed human services equitably to all. At other times, it seems to be driven by social-science arrogance and politics. At any rate the movement is with us. You can't escape it, so judge for yourself. And remember that the rich reality of culture itself is not the same as the movement.

The purpose of this brochure is to help you understand the place of diversity, especially cultural diversity, in counseling by discussing some of the principles of multicultural competence and by sharing with you how I developed as a multicultural counselor over the years. Hopefully, you will also come to understand the *Skilled Helper* model more broadly and more fully through examples of helping situations across the world. In this brochure each stage and task of the problem solving and opportunity development model, together with the communication skills needed to make it work, is illustrated by examples from outside the United States and by cross-cultural examples within the United States. While most of these examples are from my own experience, some come from students and colleagues. And don't let the citations fool you. This is not a scientific document. It is a walk through the world with the *Skilled Helper* as a lens.

Multiculturalism as a Social Agenda

Some see the multicultural and diversity movement within counseling as a social and political process aimed at seeking justice for minorities who have been treated unfairly and marginalized. For instance, the APA guidelines reflect "the continuing evolution of the study of psychology, changes in society at large, and emerging data about the different needs of particular individuals and groups historically marginalized or disenfranchised within and by psychology based on their ethnic/racial heritage and social group identity or membership" (*American Psychological Association,* May 2003, p. 377; see *The Counseling Psychologist, 32*, January 2004 for a number of articles dealing with this social-justice agenda in counseling psychology). The guidelines from the other organizations mentioned above express similar sentiments.

Although there is a debate about helpers' roles in this social justice agenda, the agenda itself is extremely important and should be embraced by all the professions and by the country itself. Of course, helpers' multicultural competencies serve all clients, including the socially disadvantaged. And the purpose of this brochure is more modest. Since counselors help individuals rather than groups and since within-group differences are often larger and more critical than between-group differences, the focus here is on the individual and how his or her culture provides one important context for the helping process. In individual helping sessions, as we shall see, culture is sometimes center stage and at other times in the wings, outweighed by other forms of diversity.

Culture Can Be at the Heart of a Problem Situation.

Culture can be a key factor for any given client struggling with a problem situation or trying to identify and develop opportunities to live a better life. Chester, a man nearing thirty whom I met in the United States, talked about the difficulty he was having in finding the right kind of life partner. He said he practically despaired of doing so. He was not your ordinary person. He had spent the first eighteen years of his life in an African country. His parents, American citizens, worked with the poor of that country. They had an interesting and certainly challenging arrangement with the children. The children lived and went to school with the poor. They lived a rather hard life. They received few privileges. Their classmates were the sons and daughters of miners. However, when they reached the age of eighteen, they went to the United States to attend college and to determine what direction their lives would take. Though on occasion they visited their parents, none of them returned to live in Africa. Eventually the parents retired to the United States.

The first eighteen years of Chester's life had a profound impact on him. He grew up with sound values. He was resilient, resourceful, and self-confident. He knew the value of hard work. But when he returned to the States, he felt somewhat like "a stranger in a strange land." For all practical purposes he was a foreigner. It was not that he felt lost. Rather he felt that many of people he met were lost. "Getting ahead" seemed to be on people's minds, while "accomplishing something" was on his. "Acquiring" seemed to be important for many, while "contributing" was important to him. Some saw him as judgmental, probably because he asked incisive questions about a world that was new to him and because he knew where he wanted to go in life and was driven by his values to go there.

Chester felt more attracted to foreign women who had come to study or work in the United States than to American women. They seemed more mature to him. In his mind they more readily understood his background and his values, perhaps because, like him, they were outsiders. But they were not here to stay. He fell in love with a Malaysian woman and would even have gone to live with her in Malaysia, but she wanted him to convert to Islam "from the heart." Outward signs would not be enough. This he could not do. So what was he to do? He talked as if there were only two stark options. Continue looking for the right partner knowing that the search was futile. Or give up, dive into his vocation completely, and not worry about a life partner.

Chester had some blind spots to contend with. Though he was not necessarily judgmental, he was not fully aware of the aura of the "right-wrong" kind of thinking that hung about him. When he talked, there was a bit of "attitude" about him. When he met American women he expected to be disappointed and inevitably was. When asked directly, he said that he wanted a spouse with values like his, one for whom "accomplishing something" was more important than "making it." He wanted to partner with someone who understood the person he had become through his upbringing and atypical personal experiences. He was an artist and wanted to use art as a way of helping young people develop a sense of the world. He wanted someone who would partner with him in all of this. A key question was this: Could he, like most of the people in the world, accept someone who came close to what he wanted? Would he be willing to engage in the kind of give and take necessary to create a mutually satisfying relationship? Or would some of the rigidities in his thinking lead him toward giving up the search? He needed to come to grips with what he wanted and what compromises he could make without violating his core values.

Culture Is Not Always at the Center of the Problem Situation.

As important as culture is, it is not always at the center of every problem situation. I counseled a worker at the World Bank who was bewildered by how differently the Bank with its Western patina did things (the

World Bank culture) from the way they were done back in his native land. In his case a set of cultural differences was a key diversity factor. He eventually left the Bank, not because he did not understand finance, but because he was uncomfortable with its social system. Here again culture was central. On the other hand, I once counseled a manager in South Korea regarding his rather acerbic management style. In this case, he differed from other managers in the same company and from the managers I met in his department. Culture was not a central factor in our discussions, but personality traits and management style were. His less-than-facilitative style was not just a by-product of Korean culture.

Consider Maria's case. She was an Argentinian citizen of Italian descent who did her undergraduate work and a master's degree in fine arts at two different Midwestern universities in the United States. While she described herself as a "hybrid, sometimes more American than American themselves," circumstances kept her outside mainstream American life. Near the end of her undergraduate studies, she was offered "the interview of a lifetime" for a big job in the arts. However, by engaging in some less-than-rational behavior the night before the interview, she blew it. This would haunt her throughout graduate school.

After finishing her graduate degree, she spent time in low-paying jobs in order to remain "legal." Her disappointment with herself lingered. Though attractive and gregarious, she failed to make many friends and certainly did not create a network of contacts that might have led to her getting a satisfactory job that could have led to a green card. She generally did not like the men who were attracted to her and those to whom she was attracted showed little interest, were already taken, or wanted to be "just friends." The small circle of close friends she did have had developed a kind of perpetual-student lifestyle. She and one close friend were "captives" of the university. As long as they continued their studies, their visas were safe, but they could work only part time in low-paying university-based jobs that led nowhere. They felt trapped with time running out. Maria and her friends spent a great deal of time talking together, complaining about their plight, criticizing the "system," staying up late, and sleeping a great deal.

When I met her, she said that she wanted to find some legal way of staying in the United States. But did she? We talked about her plight, Argentina, and the career she wanted to pursue, but she never moved on to a serious discussion and search for legitimate ways of staying in the United States. Finally, she talked about her family in Argentina and the "unfinished business" she had with them. With time running out, she decided to go back to Buenos Aires.

Maria's central issue was not culture but a need to face up to key developmental tasks of her stage of life. She was reluctant to face up to the challenges these tasks presented. In the end, her return to Argentina was her way of saying, "I'm going to face up to myself."

Personal Culture: Diversity Is Always an Issue

Recall the description of culture outlined earlier. Is there a way of applying this view of culture not just to groups, but also to individuals? I think so. I use the term "personal culture" to signify all the differences that are "packaged," so to speak, in any given individual. This includes, but is not limited to, cultural differences. But let's start with culture. Individuals manifest great differences in the ways in which they express the culture of the group to which they belong. The range of ways of thinking and of behaving is enormous. The ways in which an individual interprets and integrates the beliefs, values, norms, and behaviors of his or her culture and subcultures into his or her personal behavioral style. However, this style is not only a product of the person's culture and subcultures but of all the other diversity factors listed above—all the beliefs, values, norms, and patterns of behavior that make him or her this particular human being. I have an Egyptian friend. His personal culture describes the way *he* thinks, that is, *his* personalized package of assumptions, beliefs, values, and norms, and the way *he* does things, the patterns of behaviors that characterize him. His assumptions and beliefs about himself, others, and the world around him interact with the values that he has both inherited and acquired over time. These interactions produce his personal norms of behavior. Add to this the other diversity factors listed above and you end up with an Egyptian that is like other Egyptians and an Egyptian who is different from other Egyptians.

Not that any of this is really new. Murray and Kluckhohn (1953) made the following oft-quoted statement: "Every man is in certain respects (a) like all other men, (b) like some other men, and (c) like no other man." First, all individuals have a number of things in common. There are universals. Second, groups of individuals share characteristics that other groups do not. There are groupings. Third, each individual is in some way unique. As a counselor, I need to understand my clients in each of these three ways.

We Counsel Individuals

Whenever I am with a client, I am face to face with the complexity of his or her personal culture. Chen and Davenport (2005, p. 109) emphasized this at the end of an article on counseling Chinese Americans.

> Although cultural knowledge is essential in helping the therapist identify potential conflict areas, one must be cautious not to apply cultural information in a stereotypic manner. Because of the rich diversity within the Chinese community, it is important for mental health professionals to keep in mind that cultural difficulties (such as the degree of acculturation and racial identity), socioeconomic background, family experiences, and educational level impact each individual in a unique manner and that each Chinese American should be considered individually in the process of therapy.

Everyone has his or her own personal culture, including identical twins reared in the same way. Twins begin to diverge in personal culture shortly after birth. Here is an example of what I mean by personal culture. At a training conference in Belgium I met a doctor from Sao Paulo, Brazil, who worked for a large international pharmaceutical company. He had an Argentinian mother. He had spent a few years in the United States, but was now in charge of clinical trials in Brazil. He had come from a very poor family but had managed to work his way up the Brazilian educational, economic, and social ladders. He was urbane and charming. A number of different cultures ran through his blood. He was a Brazilian with all the basic cultural meanings of the term, but he had inherited Argentinian cultural influences from his mother. He had some residue of the culture of poverty into which he had been born and raised. He was now a member of the upper middle class and he behaved accordingly. He was a doctor and the culture of the medical profession had made its mark. Since he was a manager and a businessman, some strain of business culture ran in his veins. He was Catholic and espoused in his own liberal way traditional Catholic values. But he was a Catholic who loved to question the meaning of life and the ways of the Church. He was all of these and more, but he integrated and expressed all of them in his own way. And so when he talked about a conflict he was having with his boss and some of the related "conspiracies" to be found in the office and lab, he spoke with a voice that had many currents. However, I wasn't talking to a Brazilian or a doctor or a Catholic or a researcher or a businessman. I was talking with Renaldo and I was a learner as is any counselor who sits with a client. There were two other Brazilians at the conference, but the three of them were not Brazilians in the same way. Although I interacted with the other two only intermittently, I soon realized that the personal cultures of these three gentlemen had both similarities and great differences.

CULTURAL COMPETENCE AS PART OF DIVERSITY COMPETENCE

Cross-cultural competence refers to both the knowledge and the skills needed to relate to and communicate effectively with people from a culture that is not one's own. Over the years people have drawn up a variety of lists outlining specific competencies (La Roche & Maxie, 2003) and hefty handbooks offering "the theoretical background, practical knowledge, and training strategies needed to achieve multicultural competence" (Pope-Davis, Coleman, Liu, & Toporek, 2004) have begun to appear. In addition, there are dozens—or by now hundreds—of highly detailed research studies offering further insights into multicultural competence (see Darcy, Lee, & Tracey, 2004). Day (2005, p. 31) notes that multicultural counseling competence "is usually conceptualized as including awareness of one's own culture, biases, and values; knowledge about social and cultural influences on individuals; and skills for applying this knowledge in counseling." But it seems that there is no universal agreement as to the "right" package of multicultural competencies (Kia 'I Kitaoka, 2005).

Here is my adaptation of a list of multicultural competencies outlined and illustrated by Hansen, Pepitone-Arreola-Rockwell, and Greene (2000). This in one of dozens of views on cultural competence and was chosen somewhat randomly. I have changed the language, separated what the authors have grouped, grouped what the authors had separated, introduced ideas from different authors, added thoughts of my own, and thereby introduced my own bias. My own bias, of course, is that diversity, especially diversity as represented in personal culture, is the key concept and that culture, important as it certainly is, is one among many key diversity factors. As a counselor, I must engage the personal culture of each client I see.

- Be aware of your own personal culture, including your cultural heritage, and how you might come across to people who differ from you culturally and in a host of other ways.

- Be aware of the personal-culture biases you may have toward individuals and groups other than your own.

- As a counselor, be aware of both ways in which you are like any given individual and ways in which you differ. Both can aid or stand in the way of the helping process.

- Come to understand the values, beliefs, and worldviews of groups and individuals with whom you work.

- Come to understand how all kinds of diversity, cultural and otherwise, contribute to each client's dynamic makeup.

- Be aware of how sociopolitical influences such as poverty, oppression, stereotyping, stigmatization, discrimination, prejudice, and marginalization might have affected groups and individuals with whom you are working no matter what their culture might be. Culture is one among many targets of such abuse. Any sort of diversity— such as age, education, and disability—can become targets of these negative behaviors.

- Realize that mainstream Western psychological theory, methods of inquiry, diagnostic categories, assessment procedures, and professional practices might not fit other cultures or might need some adaptation. Be aware that some of these factors might not even fit people from Western cultures that well because of within-culture diversity and other diversity factors beyond culture.

- Get to know the basics of family structure and gender roles of groups with whom you work. Remember that there can be great differences within any given culture. Culture does not automatically mean homogeneity.

- Develop an understanding of how people in different cultures understand and deal with illness, including mental illness, and how they feel about help-seeking behavior. Remember also that people in the same culture have wide differences in this regard because of their personal cultures.

- Establish rapport with and convey empathy to clients in culturally sensitive ways. Extend this sensitivity to the personal cultures of all clients. Be especially careful not to think that people from your own culture are all alike. You are establishing rapport and expressing empathy to individuals, not cultures or other forms of diversity.

- Recognize and appreciate cultural and personal-culture differences in interactional styles and language differences, including nonverbal communication, between yourself and your clients. Remember that people in the same culture communicate and interact in a whole range of ways.

- When clients tell their stories, recognize which issues are culture-specific and which are more related to universal human experience. If a young person is having some problems with his parents, realize that having problems with parents is close to a universal experience. In the words of a Jacques Brel song, "Who is the child without complaint?" Parents aren't perfect. On the other hand, since parent-child relations differ widely from culture to culture, the specific twists of the problem are often culturally conditioned. But within-culture differences can also play a big role here.

- Design non-biased treatment interventions and plans for clients that factor in key cultural and personal-culture variables.

- Initiate and explore issues of difference between yourself and your clients when this is appropriate. Remember that culture is only one difference. In the end, your interactions with your clients are a personal culture to personal culture affair.

- Assess your own level of cross-cultural and personal-culture competence and strive to improve in all the areas outlined above.

In other words, work with your clients the way they are. Be aware of the key ways in which you differ from your clients. Appreciate clients' differences. Don't let differences interfere with the helping process. But don't feel the need to apologize for who you are.

Stuart (2004), noting that it "is easy to endorse the principle of culturally sensitive practice, it is often much harder to make it a reality" (p. 3), has written an excellent article on ways of avoiding either overvaluing or undervaluing key cultural-competence behaviors in encounters with clients. When it comes to culture, complexity is the name of the game. He notes that no one is the repository of a "pure" culture.

> Everyone belongs to multiple groups—nation, region, gender, religion, age cohort, and occupation to name a few—each of which exerts a different cultural influence that may be congruent, complementary, or in conflict with any of the others. Every influence is interpreted by each person, who decides whether and, if so, how personal beliefs should respond to each of these influences. Therefore, every individual is a unique blend of many influences. Whereas culture helps to regulate social life, specific beliefs are products of individuals' minds. Because of this complexity, it is *never* safe to infer a person's cultural orientation from knowledge of any group to which he or she is believed to belong (p. 5).

Stuart's article is an excellent introduction to the cultural-competence issue. He neither overstates nor understates the importance of culture in the helping process. He emphasizes the overarching importance of diversity beyond cultural diversity. His 12 practical suggestions for achieving multicultural, and I would add diversity, competence may not be the last word, but they constitute an excellent starting point.

As a first-generation Irish-American, I was quite aware of my cultural heritage at an early age. It was part of my identity. I even remember being critical at an early age of my own cultural counterparts. For instance, after listening to my father and his brother argue about politics, one a democrat, one a republican, I asked myself, "Why do adults try not to communicate with one another? Is this an Irish thing?" Later in life I saw a few Pinter plays. Non-communication played an important role in each of them. And I saw what Pinter wrote about dramatized over and over again in British society. I also noticed that the Irish and British engaged in non-communication in different ways. But, as I traveled the world, I came to realize that the failure to communicate is not just an Irish and British thing. Everyone does it, but in enchantingly and exasperatingly different ways.

I think the best word that describes me in encountering people in other cultures is, and I say this with all humility, "humble." An inveterate learner, not only did I right from the start spot cultural differences between myself and people from other cultures, but I also soon discovered that there were great differences from person to person within each culture. Even though I was considered an "expert" in my field in many of the places I visited around the world, I always realized that I was there as a guest and that I was a learner. I could not possibly have become an expert in all of the cultures I encountered. I knew that if the people I met thought I had nothing to offer, they would not have invited me. If they felt that I was not a learner when I arrived, they should have sent me home. I am sure that people expected me to respect their culture. But they—and I—also thought that it was important for them to respect mine. We encountered one another, it strikes me, first through *our humanity*. Then we dealt with culture and all the other many forms of diversity that rinsed through all of us, our personal cultures.

At times this was humorous. In many countries when moving from place to place in a building my hosts would almost always have me go first. Finally, after taking many wrong turns, I suggested that one of them lead the way. Once they found out that I was more comfortable with this arrangement, we all went from place to place more quickly and more comfortably—and with our cultures intact. Another humorous note was struck at dinner one evening at a conference in Northern Ireland. We were discussing the "troubles" of Northern Ireland and its relationship to both the other parts of the United Kingdom and the Republic of Ireland. As we talked, I began to wonder whether people in Northern Ireland saw themselves as Irish and not just British. Since most Americans are ultimately from somewhere else, they routinely ask people, "Where are you from?" or "What are you?" in search for the other person's ethnic and cultural identity. So I asked them, "If I went around here knocking on doors and asking people, 'What are you, what's your background?' how would they respond?" "They'd shoot you," was their reply.

Multicultural competencies are not just for helpers. Given the diversity of the society in which we live and the diversity of the world, we all need them. The Tilford Group at Kansas State University (www.ksu.edu/catl/tilford)—a research and development group consisting of inter-disciplinary faculty, administrators, and students committed to multicultural competence—has drawn up list of multicultural competencies in terms of knowledge, personal attributes, and skills they believe are needed by people to live and work in a diverse world. This list is included in the appendix at the end of this booklet.

THE SKILLED HELPER MODEL

What follows is a brief outline, stage by stage and task by task, of the outcome-focused *Skilled Helper* model. Each stage and task is briefly described and then illustrated by one or more examples with some kind of cultural twist. Then the communication skills needed to make it work are briefly noted and accompanied by some examples of cultural differences I have run into around the world. This entire section of the brochure rests on the assumption that the reader is reasonably familiar with the stages and tasks of the *Skilled Helper* model and the communication skills needed to engage clients productively in these tasks. The minimum would be that the reader has thoughtfully read Chapter One of *Essentials of Skilled Helping* or Chapter Two of the *Skilled Helper*. These chapters provide an overview of the model described and illustrated in these two books. Counseling is ultimately about outcomes. Problem situations need to be managed; opportunities need to be identified and developed.

Broadly speaking, the three stages of the model refer to three questions that clients need to ask themselves (with whatever help they need from a counselor):

I. What are the issues in my life that are causing me concern?

In responding to this question, clients tell their "stories." These stories deal with problem situations and/or undeveloped opportunities in their lives. "I've got good news and bad news. I'm about to be laid off from a job I hate."

II. What do I want (instead of what I've got)?

In responding to this question, clients discuss their goals; that is, the *outcomes* that will help them manage problem situations or develop opportunities. For instance, "I want a job that will absorb my energies, use my talents, provide some sort of career path, and provide decent support for my family."

III. What do I need to do to get what I need or want?

In answering this question, clients describe the kinds of actions, strategies, programs, and plans they need to implement in order to achieve their goals. "To get the kind of job I want, I'm going to have to review the kind of career I want, and then do some cold calling, post my resume on monster.com, canvas my friends, read the want ads that are in line with my job preferences," and so on and so forth.

From one point of view, the stages and tasks of the model are ways in which counselors can "be with" and help their clients. Being with a client means getting to know and understand the dimensions of the personal culture of the client that are relevant to the client's problem situation. Not every client needs extensive help with every stage and task of the model. Good counselors continually develop a deeper understanding of the personal culture of their clients and of their clients' problem situations and help them choose the mix of tasks that best fit them and the issues with which they are struggling—all of this at the service of problem situations "better managed" and opportunities for a better life "identified and developed."

The *Skilled Helper* has been used successfully across a wide range of cultures for some time. One reason for this is that problem management and opportunity development, which are at the heart of the helping process outlined here, are human universals. At least that is my conviction. However, the fact that it has been welcomed in many cultures does not belie its Western roots.

I wondered about this during a lecture tour that included three Muslim universities in Malaysia. First of all, I had some of my stereotypes about the place of Muslim women in Muslim cultures blasted away. For instance, the women instructors I met and who shepherded me around were all gracious, strong, and efficient. I watched one woman in action and thought to myself, "She could easily run a U.S. company, but then why would she want to!" She was an excellent communicator. She knew exactly what would help the students. She was very well organized. We began on time. The students were well prepared, participated very well, and asked incisive questions. We ended on time. I wondered about my Western roots and the universality of my helping and management models in the afternoon when I gave a talk on management in educational institutions to faculty members of the same university. As I looked at the expressions on the faces of the men and women there, I could almost feel their reservation, "Well, this is one way of looking at the world. We'll think about it." I was ready to attribute what I saw as resistance to the culture gap between East and West and between

Christian and Muslim, but then I remembered seeing the same look on the faces of faculty members during a similar talk I gave at Notre Dame. "There may well be a culture and religious gap here," I said to myself, "but it is also the resistance of university teachers listening critically to one of their own." It wasn't just Asian or Muslim culture. It was university culture. Indeed, when talking to the students of these universities, I saw mostly intellectual curiosity and enthusiasm.

Stage I—Help Clients Explore Their Concerns

The three tasks of Stage I are pretty straightforward:

- Help clients tell their *stories* — "Here's what's going on in my life and here's how it bothers me."
- Help clients identify and deal with *blind spots* that keep them from seeing their problem situations and unused opportunities clearly.
- Help clients work on the *right things*, things that will get them where they want and need to go.

As we shall see, these three tasks are useful throughout the helping process. For instance, blind spots and the need for new perspectives are not limited to the storytelling part of the helping process.

Task 1: Help Clients Tell Their Stories.

Clients come with stories to tell. Counselors help them tell these stories. The skills of active listening, communicating understanding, and helping clients clarify their messages are essential. Counselors add value when they help clients stay focused, keep them from rambling, and see clients' stories as just the beginning of the journey toward problem-managing and opportunity-developing outcomes. Clients need to act in order to manage problem situations and develop opportunities. So looking for ways to help clients act on their own behalf starts with the stories themselves.

Often enough not all the story or relevant facts come out at once. Key dimensions of stories tend to "dribble out." And so, a discussion of problem-managing goals, a Stage-II activity, can be influenced by new data, a different twist, or a further elaboration of the original story. A client might be well into grappling with one problem when he or she brings up another. The new problem is often the more important problem.

Here is an example of my helping a client in China tell his story. At the beginning of a Skilled-Helper conference in China I was told by the translator how things worked in Chinese classes. I was to give a lecture. The students would copy down what I said. There would be few questions because I was the expert. I looked out at my audience, about 60 Chinese students and counseling professionals with an age range from 20 to at least 60 and said to myself, "This I can't do." I dutifully lectured for a while, but then I violated all the rules. I wanted to illustrate what the first task in the helping process might look like. So I asked the participants to write down a few things they were struggling with in their lives. Then I said, "Pick one of the issues that you might be willing to talk about in class." Finally, I asked for a volunteer. The volunteer was to stand in front of the class while I stood at the rear. My job as a counselor was to help him or her tell his or her story.

A young man volunteered. Here are the highlights of his story which I helped him tell by using active listening, sharing empathic understanding, asking open-ended questions, and other methods to help him tell his story as clearly as possible. Helping clients tell their stories clearly helps them understand themselves and their problem situations better. He was a university student in a social studies program with an emphasis on counseling. He talked a bit about some of the problems he had being a student. But then, perhaps because he had gained a bit more confidence and courage, he launched into a new problem. He said he was having second thoughts about choosing a career in social studies. He knew that if he did pursue a career in counseling, he would add value to society but, he added boldly, he would not make much money. A number of his friends had gone into "commerce." Some of them were doing very well. Of course, this brought up all sorts of issues—the new direction China seemed to be taking, a different kind of economy, the pitfalls of capitalism, being a traitor to the nation's social agenda, and related concerns. But, despite all this, he felt that he still would like to try the business world. During our dialogue the participants from time to time would gasp, whoop, and holler at what he was saying. It was good theater and his stage presence was great.

During these sessions we all learned that the culture of classroom interaction as outlined by the translator was not set in stone. By and large the younger members of the group were up for trying new things. And trying new things did not make them any less Chinese. They thought it made them more Chinese in a China that was moving out into the world. All of this took place in a spirit of good humor. A representative of the group from Schezwan, a province known for its hot dishes, raised his hand and said, "We from Schezwan are delighted by the humor you use while teaching. We see ourselves as the best humorists in China and we appreciate it." I replied, "So Schezwan is spicy in more ways than one." The laughter of the group told me something important. Some of the participants laughed right after a humorous remark. Others laughed only after hearing what the translator had to say. Only those who knew English "well enough" were supposed to attend the conference. But "well enough" was proving to be quite an elastic term. So I took pains to put things as simply as possible, not because I thought the attendees were not intelligent. Rather, I wanted to make the translator's job as easy as possible. Help also came in an unexpected way. When those who did know English well thought that the translator did not get it right, they spoke up and told him so on the spot. Another bit of culture.

Task 2: Help Clients Move Beyond Blind Spots to New Perspectives.

Blind spots are important things that people fail to see or refuse to see as they move through the helping process. There is no individual immunity to blind spots. Ideally, a challenged blind spot turns into a "new perspective" that helps clients see a problem situation or an undeveloped opportunity in a different light. Blind spots can pop up throughout the helping process. Because of blind spots clients can see problems but overlook opportunities, fixate too soon on a goal instead of exploring possibilities first, avoid good strategies to accomplish goals because they instinctively know these strategies will entail a lot of work, or fail to review obstacles to otherwise good plans. So helping clients challenge their blind spots as they tell their stories, set problem-managing goals, figure out ways of achieving these goals, and implement their plans is a very important part of the helping process.

Since cultures, including personal cultures, provide lenses through which people see the world, they are breeding grounds for blind spots. Often enough the implications of different cultural viewpoints need to be reviewed and challenged. For instance, many people I met in other countries expressed concern, dismay, and even disgust for what they saw as an overly permissive culture in the United States. This not only sensitized me to this dimension of my own culture, but led me to share some of their concerns. One of the broad beliefs of a permissive culture is that "I can do anything I want." This principle is seen as or felt to be liberating, whereas it is often dangerous and self-destructive. Some of the clients I have met not only espoused this principle, but added to it: "I can do anything I want, but, of course, if I get into trouble then you have to save me." Again, the principle may never be voiced, but it can and does drive behavior that is both self-destructive and inimical to the interests of society.

Another source of cultural blind spots is the fact that not every practice within a culture remains useful forever. Sometimes it is not useful in the first place. When in the 1880s one of the leaders of the Maoris in New Zealand, now a group with real social troubles, was asked why his people still made sorties to neighboring islands, killed the men, and brought back the women, he answered, "It is our custom." Not a very useful custom (see Diamond, 1997). The world changes and some cultural practices become outmoded, even though they live on. They no longer serve the purposes of the group. They become cultural blind spots. Years ago a noted Chicago-based organizer was working with a group of native American Indians. To say the least, he was sympathetic to their cause. He listened carefully as they talked about some of their attempts to deal with social and economic problems. At one point he asked, "Why are you doing it that way?" They paused, then answered, "Because it is the Indian way." He retorted, "What do you mean, the Indian way! Do you mean that being ineffective is the Indian way? What you're doing isn't working and couldn't work. How can it be the Indian way?" He was not putting down Indian culture. He was raising the possibility that they had become victims of a cultural blind spot. He was following one of Stuart's (2004) cultural-competence directives: "Respect clients' beliefs, but attempt to change them when necessary" (p. 6). Stuart goes on to say, "Empathic therapists see the world from the client's perspective, but they do not necessarily accept everything in the client's view as healthy" (p. 8).

Cultural understanding should be a two-way street. One source of cultural blind spots is the assumption that one's own culture takes precedence over others. The practices of culture X might well collide with the practices of culture Y. What do we do when cultures collide, as they often do? If an Ethiopian meets a Thai at a conference in Russia, how are they to interact? When we come face to face with another culture or subculture, we are all learners. People learn a great deal about their own cultures when it begins to rub up against another culture. The Thai and the Ethiopian should both be learners with respect to their own cultures and the Russian culture in which they temporarily exist. This is also true, of course, of personal cultures. Unfortunately, "rubbing up against" often produces conflict instead of learning. As to the Thai and Ethiopian, more important than culture is their basic humanity. That is, their interactions should be guided by mutual respect and respect for their Russian hosts. Most of the multicultural literature in counseling focuses on the obligations of the counselor. What about the client? If counseling is a partnership, a therapeutic alliance, what can a client do to make a cross-cultural partnership work? The current literature is silent on this issue.

Of course, not all blind spots in multicultural situations are cultural. Consider this example. The East Asian head of one of the research departments in an international agricultural center located in the Philippines had a blind spot, it seemed to me, that affected both his work and his own sense of well being. His problem situation was that he was being asked to take on more and more "administrative" tasks which diverted him from his real love, that is, research. In his mind administration or management was in many ways inimical to research. In truth some of these international research centers were poorly managed. And so donor groups were forcing them to become more "businesslike," developing strategic plans, setting clear goals, focusing more directly on results in farmers' fields, and the like. The donor groups made it clear that they were not funding "pure" research. That was the work of universities and these agricultural research centers were not meant to be like universities. Their aim was much more practical—helping feed the people in developing countries.

I searched for ways of helping him see that what he called "administration," often with a curled lip, could really be transformed into effective management and, even better, leadership. Of course, I sympathized with him because as a management consultant I had come across a lot of bureaucratic nonsense in "administration." He needed to move beyond his prejudices regarding administration and focus on best practice in management and leadership. In one coaching session I asked him, "What's more important here: to do good research, or to make sure that the right kind of good research; that is, the kind of research that makes farmers' fields more productive, takes place?" He looked caught. "Well," he said, "if you put it *that* way, I'd have to say the latter." I said, "Well, that's what I mean by effective management and strategic leadership instead of mere administration." We went on to have a much more fruitful conversation. Ultimately, he went on the become Director General of the center and win an international food prize. I am not fool enough to think that my intervention led to all that. But it did lead to his rethinking his "administrative" role at the center.

Some blind spots are found within a culture rather than between cultures. I was at a conference in Guangzhou, China attended by Chinese entrepreneurs and Chinese managers who worked for large international corporations such as General Electric. About 30 managers attended, half entrepreneurs running their own businesses, and half managers from larger corporations. These managers came together because they thought they could learn from one another. The entrepreneurs were especially eager to learn how to manage and lead better in order to "make it big." Some of the entrepreneurs were quite successful and understandably prone to the kind of arrogance that can accompany business success.

One of the entrepreneurs asked to meet with me after the conference was over. It proved to be a coaching/counseling session. His problem was this: His top managers kept quitting. When asked about his management style, he replied that it was the style of an owner, not a manager. He told his workers what to do and then made sure that they did it. In his eyes that was the Chinese way, at least for an owner. Obviously his style was not working with his best people. He asked me what he should do to retain his best people. "What do those who quit end up doing—I mean the best ones?" I asked. "A lot of them start their own businesses," he said. "How well do they do?" I asked. "Some of them do very well, sometimes even better than I do." "If you still had all that managerial talent, what would your business be like?" I asked. He thought for a moment. "Booming," he said, "booming." "So you actually *need* them, is that right?" He paused, then said, "I never thought of it that way. I thought they needed me because I owned the business and they needed work." We went on to discuss what his best managers wanted. Some obviously wanted to run their own business. His business was practice for them. Others would like a *piece* of his business, however that might be defined. Anyway, our conversation helped him rethink his approach to management.

Here is another example of a blind spot within a culture. When I first went to China I was told how important relationships were. Things got done through relationships. At the time I did not catch the cultural meaning of the term. I was enlightened, then, some years later when I intervened in a conversation between two managers, one older, one younger. They had been discussing the need for leadership in business. The older manager was emphasizing the importance of relationships. The younger one reacted violently. "No," he said, "that's the old China. That doesn't make sense in the new China." They began to argue. I intervened by asking what each meant by the word "relationship." The younger manager thought that the older manager was talking about relationships in a "business as usual" way. That is, you get things done through political affiliations, by knowing the right people, by warming up to the right people, by getting on the right side of the right people, and so forth. The older manager had actually meant something quite different. He meant good relationships among people who have to work together to get things done. He meant cooperation and collaboration. The younger man's blind spot lay in his assumption that all older managers meant relationship in some corrupt political sense, meaning this: They were actually talking about more or less the same thing.

The "more or less" part proved to be quite interesting. I suggested that in an ideal world the mutual respect and collaboration approach to relationships made excellent sense. However, since we don't live in an ideal world, the political or power dimension of relationships was not going to disappear either on the world scene, in businesses, or in social relationships. Human interactions have a "shadow side." A failure to understand and deal with the political and even the more sinister dimensions of relationships would be counterproductive. Understanding the shadow side and developing the ability to spot and deal with it was important even for those who preferred and worked toward establishing relationships based on mutual respect, honesty, and collaboration. They decided that "moral, smart, savvy, and wise" (or whatever their Chinese equivalents were) all had a place in business transactions. At any rate, we all learned something from the discussion. I learned that there were probably a lot of different subcultures within the Chinese business culture and clashes of cultures could be expected.

Task 3: Leverage—Help Clients Work on the Right Things.

Clients often show up with more than one problem. Or the problem situation they have is complex with many twists and turns. Task 3 is about the "economics" of helping. Helping should not be a waste of the client's time nor the helper's time. Therefore, counselors should help their clients work on the "right things." For instance, if Yolande is having sleepless nights and has a messy financial life because of overspending on her several credit cards, it probably makes sense to focus principally on her mismanaged finances. Mismanaged finances have a way of keeping people up at night. On the other hand, helping her sleep soundly in spite of her financial woes might well lead to disaster. Helping clients identify and work on key issues within problem situations is central to the economics of helping. It sometimes involves some kind of triage system—critical issues, important issues, useful issues, and nice issues in descending order.

In the following example, culture played a central role in determining the most important part of the problem situation. A young Jordanian at a university in the United Kingdom approached me warily one day. He said that he'd like to talk about something, but not "right away." I said, "Whenever you're ready." Weeks went by. But I took him at his word and did not remind him of his request. One day he asked if we could have a cup of coffee. Over coffee he said that he was glad that I had not pushed him to talk. He talked about many things. I found out that he was not a Jordanian, but a Palestinian whose family had fled to Jordan during the war. He wandered through a number of different concerns. He worried a lot about his family. For all its bustle, he found London a lonely place. He mentioned he was gay but not involved with anyone. He had spells of depression, but managed to get by. Finances were shaky, but he also put this in the "getting by" category. He was studying business, but this did not excite him very much. The also expressed concern about the "region" and wondered how the Palestinian question could ever be resolved. He was not a militant, but he felt guilty because he wasn't "doing anything" for the region. He mentioned that he had two brothers and a sister. He was the oldest. He said that he did not want to go back to live in the region, but did not know what his other options were. He thought they were slim. Anyway, leaving the region was "copping out."

He had a lot of problems. Or rather his problem situation was complex. Where to start? What to work on? What was he struggling with most? What issue, if resolved, would take care of other issues or at least provide him with the energy needed to attack other facets of the problem situation? Here is what I discovered during a couple of meetings in which I tried to understand him in context and probe for what was most important to him.

His core problem had a very cultural twist. In many ways his family was his life. As the eldest son he was expected to marry and have children. As the eldest he was supposed to be the first to marry and the one who would take possession of the family home. Being gay was anathema. If you were gay, you got married anyway, had children, and continued the family tradition. He loved his family, but he felt he could not go home. He felt he would disgrace his family and his community. He was depressed because he saw no way out.

It would have been easy for someone to suppose that the "real" problem was being gay, but in fact he had come to terms with being gay. But he wanted help in figuring out how not to disappoint or disgrace the people he loved: For him getting married would be living a lie. He wanted his brothers to get married and carry on the family traditions. His other concerns were real, but this was central. "How can I establish a base outside Jordan that will allow me to do my part in providing for their welfare, visit them, and work for the betterment of the region?"

Stage II—Help Clients Craft A Better Future

The heart of counseling is not found in merely helping clients discuss problem situations and unused opportunities but in helping them identify and develop or "craft" a better future. "What do you want instead of what you've got?" Since it is a *solution-focused* question, it is at the heart of helping. Stage II has three tasks:

- Help clients discuss *possibilities* for a better future.
- Help clients make *meaningful choices* and set *goals*.
- Help clients explore what real *commitment* to these choices looks like.

Of course, possibilities, choices, and commitment made at this stage can be reviewed later, both in Stage III and "out there" when clients are implementing their plans.

Task 1: Help Clients Explore Possibilities for a Better Future.

Often clients come for counseling locked into their problem situations. They talk a lot about the past and their present misery, but not the future. Whatever imagination they might have seems to disappear in light of adversity. Therefore, helping clients move away from a problem focus to a solution focus is essential. It's as if the helper says, "All right, I see your misery. What would things look like if they looked better, even just a little better?"

In this example, the person with the problem situation does most of the work himself. Culture plays a key role in his struggle. One bright Saturday morning I took a taxi to O'Hare airport to catch a flight to, I forget where…somewhere. But I have not forgotten the driver, a young man born in Pakistan but now living and working in the United States. I knew he was a permanent resident, but I'm not sure whether he had become a citizen. He was quite talkative and upbeat, almost breezy. "How's life going?" I asked early in the ride. He remained upbeat but became more serious. He talked about his parents. They had come over from Pakistan, but, he said, they had never really adjusted to life in the United States. They missed the community life they had back home. He had made sure that they were quite comfortable materially, but they became restless and on edge. Recently they decided to go back home.

When I asked them how their return would affect him, he paused and launched into the rest of the story. He had a girlfriend, a woman he wanted to marry. But his parents had just told him that she was not right for him. They disapproved of her. He explained how important family and family relations are in his culture.

"So what does your future look like?" I asked. "Frankly, I don't know," he said, "or rather there are things I know and things I don't." He went on to say that he would not marry a woman unless his parents approved of her. He was sure he would marry, but he wasn't sure where he would find the right partner or where he would live. "Staying single is not an option. I could go back to Pakistan and settle down. I could go back, find a wife, and then come back here." "What do your friends here say about all that?" I asked as an indirect way of finding out how important cultural imperatives were to him. "Well, you know," he replied, "some think I'm old-fashioned; some think I'm crazy. My best American friend says that he does not share my beliefs and principles but that he sure admires the way I stick to them. To tell you the truth, breaking off my relationship with my girlfriend was not easy, but I knew what I had to do." He went on to discuss a couple of side businesses he was involved in with a couple of partners. His future was uncertain, but it was not bleak. He had options. There were opportunities out there, wherever he was. As I paid the fare, he said, "This was a great conversation, but hey, don't feel sorry for me. I'm going to make it just fine."

As mentioned earlier, blind spots can happen at anywhere in the helping process. Helping clients identify and deal with them whenever and wherever they pop up is almost invariably useful. Here is an example of a Stage-II blind spot. I met a young Peruvian woman at a conference in Buenos Aires. I talked with her a few times. She had done her undergraduate studies in the United States. Now that she had returned to Peru, she was uncertain about her future. She had many different interests. Each time we talked, no matter what the topic—for instance, such different issues as the power of the arts, social change in Peru or, more widely, in South America, or what her family life was like—she brought up directly or indirectly the possibility of teaching at a university. I began to wonder where this theme was coming from and whether she was going to end up locking herself into a teaching career prematurely. To tell the truth, it struck me that she was an unlikely candidate for a university career. But it was her life, not mine.

In our last chat, after she brought up the idea of a university career once more, I said something like this: "A frequent theme in our conversations is the possibility of a university career. I know you like the way I've lived out my university career. I've managed to do it 'my way.' But what I've done is hardly typical." She recounted some of the things she liked about my career. I replied, "Let me share an idea with you. It may be way off track. You'll have to tell me. When you talk about a university career, the term seems to be, well, almost a metaphor. For instance, it seems to be a metaphor for your deep interest in ideas. It seems to be a metaphor for social change through the influence you could have on your students. It could even be a metaphor for the success you'd like to have and the approbation you would like to have from your family. Maybe it's a metaphor for the generativity that you would like to pervade your life. You seem to value all these things, but it strikes me that there are a range of careers in which they could be embedded. The university might be one." She looked almost startled for a few moments and then we launched into a discussion of the kinds of careers that could embed such values and be carved out in Peru or wherever. She moved back from making a covert, and perhaps premature, decision and began talking about possibilities.

Task 2: Help Clients Design and Shape Problem-Managing Goals.

Once clients review a number of possibilities for a better future, they need to make choices. They need to choose goals that will help them manage the critical dimensions of their problem situations and develop targeted opportunities. Ideally, the chosen goal or goals will be specific, substantive, realistic, sustainable, prudent, and flexible. Counselors can help clients choose or design or "craft" goals with these qualities. In some cases they take time to evolve. Effective counselors help clients craft realistic goals that are in keeping with clients' values.

On a trip to South Africa during which I gave a number of lectures and workshops at various universities, I met an Afrikaner who was in the middle of making one of the more important decisions of his life. This was before the post-apartheid elections that made Nelson Mandela the country's leader. At the time there was a great deal of tension in the country. No one knew what would happen after the elections. Many feared that the country would be thrown into chaos. He was understandably worried about the security of his family. Some of his friends had decided to leave the country. He too had been tempted to do so. However, he felt that leaving would almost be an admission of guilt. This flew in the face of the fact that he had spent his university career working with black communities. He was not a racist, but his involvement with the black community had been personal and social rather than political. Maybe he should have spoken out more, he mused. He explored the reasons for leaving the country for a few minutes. Then he said, "Our family has lived here for almost three hundred years. This is the only land we know. It doesn't make sense to leave our roots, our home, our homeland." He went on to list the reasons for staying. I asked him to describe what staying would look like. This was difficult since the future was so uncertain. He soon realized that staying itself was not enough, that "staying" itself needed more definition. We spent the rest of our time together exploring *how* he would like to stay, how he might contribute through his personal life, his community, and the university to whatever new South Africa was to emerge after the elections. In the end, he did stay. And I assume that he is still doing his best to be a contributor.

In the following case, the problem situation is conditioned by social, political, and cultural factors. Years ago on my way to Australia I would stop in Fiji to see Kevin, an Irish priest friend of mine. On my second visit he said he wanted to talk through an issue that was bothering him. I had about three sessions with him. The problem situation was complicated. Fiji, though often portrayed as one of those beautiful South Pacific Isles, is actually comprised of some 300 islands spread out over about 250,000 square miles of ocean. Fiji is in many ways idyllic, but it has had its share of social and economic conflict. Historical events have created a demographic divide. Indigenous or native Fijians comprise about half the population. Because of historical covenants designed to protect the land rights of the natives, they exercise control over about 80 percent of the land. There is also a very large East Indian minority population. East Indians tend to be aggressive business people. Through leases they control the economic productivity of about 90 percent of the land. Native Fijians tend to be Christians. East Indian Fijians are mainly Hindus. Some are Muslims. All of this constitutes a formula for political, social, and economic conflict and recently there has been a great deal of political turmoil. This oversimplified thumbnail view of Fiji was the context of our conversations.

The members of Kevin's congregation, native Fijian Catholics, another minority group, were not aggressive business people. He feared that they could be victimized by the political turmoil that he thought was imminent. Although he had lived and worked in Fiji for over 30 years, he realized that he was an outsider, not pastorally or socially, but politically. He was not sure what role he could play in helping the members of his congregation find ways of living productively in their country and of preparing themselves from turmoil that seemed inevitable. He knew he was not a crusader, but what role made sense? He needed to craft a role for himself.

In our conversations he went back to his roots—the messages of peace and justice found in the Gospels. He knew many East Indian Fijians and realized that he had to take their rights into consideration, too. While there were many views of justice in Fiji, he believed that his role was to help people guide their interventions by moral and not just political and social principles. He also believed that he should work in the background, providing counsel and pastoral encouragement to the Fijian leaders in his congregation.

Task 3: Help Clients Commit to a Better Future.

It is one thing to set a goal, another to carry it out and seize its benefits. Therefore, counselors can help clients not only set goals but also examine the kind of commitment needed to accomplish them and to explore the kind of resources available to help them commit. One way to help clients get a feeling for what day-to-day commitment is like is to get them to picture themselves pursuing their goals out there is their everyday lives. Clients need to know how to spot signs that they are moving ahead or slacking off. They also need to know what obstacles might pop up and have plans to deal with them. Commitment is a process, not an event. If it were an event, then New Year's resolutions would transform our lives. "I really want to get into better physical shape this year; I'm not going to let anything get in my way" is a nice sentiment. But it is a form of event commitment. The day-to-day struggle with diet and exercise demands operational commitment.

Liang, a Chinese instructor in psychology, had a number of problems. He was officious and demanding. In conversations he would interrupt people to correct what they were saying. He complained a lot. People tended to avoid him. I found it difficult to be with him. He was his own worst enemy because his interpersonal style blinded people to the fact that he was really a decent and generous person. He also had no illusions about himself. He knew he had strengths, but he knew his effectiveness was diminished because of his style. He had some kind of master's degree, but he knew he would not go very far without a doctorate. He said he wanted to go to the United States to get a degree, but he needed both professional and financial support. I thought he did not have a chance. I could not see him as a practicing clinical psychologist. I wrote a letter of recommendation and tried my best to give a balanced picture. Liang was persistent. He tried a number of schools and was finally accepted into a decent program. He got the financial aid he needed for the first year. But one of his strengths was that he was a very hard worker. And he was meticulous. Both of these qualities were appreciated by the researchers at the university. He worked hard. He did his best to keep less attractive behaviors in check. He got the degree. He also learned—or perhaps it was just a case of having his suspicions confirmed—that clinical practice was not for him. He was a researcher. Sometimes I felt bad because I had not responded to him very well. When he said that if we had not met in China, he would never have started on his journey, I did not feel much better. I had never "leveled" with him. He deserved better from me. It may well be that meeting me was the spark that got him started, but the commitment, the persistence, the drive to succeed, that was all inside him. He responded to opportunities. Then he went on to create further opportunities. When the big one arrived, he was ready. And he gave it his all.

Often enough a major obstacle to commitment is the web of competing agendas in clients' lives. I ran into this phenomenon on a trip to Southeast Asia not long ago. One of the professors I met was from Morocco. He brought up the possibility of translating the *Skilled Helper* into Arabic. We went on to discuss a range of projects in which he was involved. It eventually became clear to me that he had too many competing projects. I doubted that the translation project would get to the top of the list soon. I asked him to contact me when he was ready to talk about it. I am still waiting. We all face competing agendas in our lives. So do our clients. Helping clients review the competition their goals will face from the other tasks of life adds to the realism of the helping process.

I also realized that a translation into Arabic would not be an easy task because of the great diversity among Arabic-speaking peoples. Many of the examples would have to be changed because of religious and other cultural considerations. It seemed to me that, while the core of the problem-management model itself had a kind of universal application, different versions of the book would be needed in different parts of the Arabic-speaking world.

Stage III: Help Clients Develop Plans

Stage II deals with helping clients decide *what* they want and making an initial commitment to the goals they choose. Stage III deals with helping them with *how* they are going to pursue and achieve their goals—the strategies and actions that will turn goal choice into goal pursuit and goal accomplishment. Stage III has three tasks:

- Help clients review a number of *different* ways of pursuing and achieving their goals.
- Help clients pick the strategies that *best fit* their personal cultures.
- Help clients but these strategies together in a coherent and realistic *plan*.

Planning for many people is such a bore. But the research suggests that planning, done correctly, is a powerful tool for change. Your job is to help clients turn it into the powerful tool it is meant to be.

Task 1: Help Clients Develop Strategies to Achieve Goals.

Clients might know what they want but are less sure about how to go about getting what they need and want. I might know that I need to be less critical and control my anger, but knowing this does not automatically give me a plan or program to achieve this goal. Helping clients identify strategies and programs to achieve their goals can add great value.

The following experience took place while I was visiting a family in Japan. The mother of the family I was staying with, let us call her Amiko, ran the household and taught English on the side. To the children I was a cultural anomaly. They laughed at my attempts to use chopsticks and were amazed or horrified by the way I mixed all sorts of ingredients into the rice I ate. Amiko had one son in full-flight preparation for the exams that would determine what kind of university placement he would get and a second son about to engage in that arduous work. Her neighbor had a son who was also preparing for the university placement exams, but unlike Amiko's son, he seemed to be breaking down under the strain. When she heard that "a noted psychologist" was staying next door, she wondered if she could get some "advice." I don't tend to give advice even in my own culture, much less in others. And I don't speak Japanese. But, since Amiko wanted to be helpful, we came up with the following plan.

The neighbor's goal was clear. She wanted her son to pass the exams, get a decent university placement, and retain his physical and mental health. What could she and her son do to achieve the third of these three goals? Amiko's son was under the same pressure, but he was doing fine. It struck me that Amiko could share some of her parental wisdom with her neighbor. She did not want to do this directly because she thought that would be interfering with her neighbor's life and could be taken as a veiled criticism of the way her neighbor was running her household. This was a very big cultural no-no. "In a country where harmony is prized, direct criticism, even the constructive kind, is still often unacceptable" (*Economist*, January 17, 2004, 36). So before sitting down with her neighbor, I interviewed Amiko and her husband about what they thought contributed to their son's ability to cope, and cope quite well, with the agony of preparing for exams. The items on their list were filled with common sense. For instance, she and her husband had made it clear that their love for their son did not depend on his passing the exams and getting the best university placement. They provided support during his agony. They recognized his moods. When he was down, they offered encouragement, but they did not overdo it. They made a special effort to be calm when talking with him and they made sure that their own interactions created an upbeat atmosphere. They provided breaks for him when he seemed to be pushing himself too hard. They presented a unified front; their son got the same messages from both mother and father. They made sure that he had a quiet place to study and warded off attempts by his brother and sister to invade his space at inopportune times. Of course, they had never thought of all of this as a plan. Rather they instinctively did what they thought to be right. It also helped that their son's motivation was high; he was eager to get the best possible placement for himself.

Since Amiko's neighbor did not speak English, Amiko had to translate. When the three of us got together, I helped Amiko share the parental wisdom that she and her husband had been exercising. She told her neighbor that Dr. Egan did not give advice but that he would share with her some of the kinds of things that might benefit her son. During the conversation Amiko and I helped each other present the various options in as constructive a way as possible. At any rate, we navigated our way through the session. Her neighbor left quite pleased. She had not made her choices yet, but she had a range of possibilities. And she did not feel that anyone was preaching to her or blaming her for not being a good parent.

Task 2: Help Clients Choose Strategies That Are Best for Them.

Some clients choose the right goals to manage problem situations and develop unused opportunities, but they choose the wrong strategies for achieving them. If clients brainstorm strategies for achieving their goals, they then need to choose wisely from the pool. Helping clients make wise choices can add great value. Ideally, Amiko's neighbor went home and discussed what she had heard with her husband. Together they discussed which strategies might be most useful for their son. I never did find out what happened to the boy.

In the following example, national culture did not seem to play a big role. Asim, an Egyptian student seeking Microsoft Certification, chose to enroll in an expensive pay-up-front training program in the United States even though his finances were shaky. The program proved to be fly-by-night and the "best" instructor a dud. He talked through his plight with a counselor at a local university and chose instead a rigorous self-study course coupled with a series of exams. The counselor thought the self-study program made sense because of Asim's intelligence and his determination to succeed. When he passed the first two certification exams (there were seven in all), Asim became even more enthusiastic about the course of action he had chosen and more organized in pursuing it. Perhaps because of ethnic culture or personal culture—hard to tell which—Asim was reluctant to go to the director of the training program and ask for at least some of his money back. When he did go back, the director said no. He thought that Asim would not put up much of a fight. But the counselor put Asim in touch with a legal aid office. Through simple arbitration he was able to recover some of the money he had invested in the aborted program.

In the next example, the choice of goals and strategies is culturally and individually conditioned. Annan, a Palestinian with an Israeli passport and permanent residency in the United States, wanted a wife. He dated women in the United States but said that he found none of them "suitable." A minister in his Church realized that it was impossible to determine how to help him find a wife without determining what "suitable" meant. It turned out that "suitable" in Annan's mind meant a woman with all the "back-home" virtues who would be willing to live as a back-home person in a completely different culture. Obviously this narrowed the possibilities. Annan returned to Palestine for a visit to begin the process of finding a wife. After a second and third visit, he finally realized that women with the right back-home qualities wanted to stay back home. The minister encouraged Annan to be transparent in his search. He should "come clean" with the woman of his choice and with her parents and family: about what he wanted, about what life would be like in the United States for her, and about the advantages and disadvantages of such an arrangement. Annan is currently doing precisely this.

Task 3: Help Clients Develop and Kick Start Plans.

Most problem situations take time to develop and, understandably, time to resolve. Research shows that making plans can add great value to the problem-management and opportunity-development process. Planning makes time an ally rather than an enemy. However, the trouble with plans is this—clients make them and then discard them. Perhaps "discard" is not the right word. Rather, plans tend to fade. When I was in Botswana, I was told that the waters that flow into the Okavango Delta, creating a large wetland teeming with animal and plant life, eventually disappear into the surrounding Kalahari Desert. No seems to know exactly how. Does this sound like any of your plans? "Where are my

New Year's resolutions?" I might ask myself in July. "They're in the Kalahari," could well be my answer. Plans are bedeviled by entropy; their energy dissipates over time. Therefore, in discussing plans with clients, it is also necessary to discuss the kinds of obstacles that can send the plans to the Kalahari. This is a continuation of the commitment process discussed in Stage II.

The next example takes place in the World Bank. But first let me give you a bit of background. I spent seven years as a consultant in management development at the World Bank. From a cultural point of view, the Bank is a very interesting place with more than 100 nationalities working there. On the surface there seemed to be relatively few conflicts. Perhaps this was due to the fact that most of the professionals had advanced degrees from Western universities so that there was a Western "patina" on most transactions. Or it could be that engagement in and loyalty to a particular profession helped people "speak a common language," as it were. Or it could be that the culture of the Bank itself demanded a certain formality and civility that kept the peace. There were conflicts, of course, and these were aided and abetted by what I called a culture of criticism and perfectionism at the Bank. But people found ways of keeping the peace. I had been around a few months when I noticed that most documents I read had the word "draft" on the cover. I asked whether there were any finished documents. I was told that most documents were finished shortly before they were to be presented to the board or to some committee that would pass on the feasibility of what was being proposed. Strategies like this helped take the edge off the culture of perfectionism and criticism.

Often enough simple plans are better than complicated ones. I was a coach and consultant to Sean, the man who ran the management development unit at the Bank. Two of his assistants, one from Kenya, the other from the Philippines, were in constant conflict. They both pleaded their cases with Sean. He found himself listening to endless recriminations and tried to be fair. Since their conflict was disrupting the work of the department, the goal was harmony or at least the kind of civility that allow them to collaborate in getting their work done. The goal was right, but his haphazard plan for them—a string of suggestions on what to do and how to get along with each other—was getting nowhere. I suggested to him that his assistants had put the monkey on his back and that we should talk about a way of putting the monkey back where it belonged. The plan that emerged was a simple one. He told them that they had two weeks to put their house in order. He would not listen to their endless stories, nor would he provide any more suggestions. He expected them to come up with a way of establishing harmony. If they did not, they would both be gone. Within two weeks civility, if not harmony, had been restored. They cooperated. The work got done. There were no more closed-door sessions with either of them.

It is not always possible to come up with a plan for achieving goals all at once, "up front." The plan can start in the counseling process and then grow, change, and develop over time. Cheung, a Chinese man, wanted to work out a career in psychology. At the time I met him, he was married, had one child, and was studying counseling psychology at a university in the United States. His wife, child, and extended family were back in Shanghai. He had taught psychology in Shanghai, but now he wanted to prepare himself to take a leadership role in the field in China. I met him as he was finishing a master's degree in the Midwest and getting ready to continue doctoral studies on the East Coast. He was quite curious about my "other" career—teaching, writing about, and consulting in management. The more he learned about the management side of the world, the more he wanted to combine a career in psychology and management consulting. He began to include courses with a business focus in his program. I mentioned his name to a friend of mine in a human resources consulting firm. My friend said that he would like to talk with him since the part of the practice he was running would soon be expanding to Southeast Asia and, eventually, to China. To make a longer story short, he ended up taking a job in my friend's practice. I would see him from time to time in the role of mentor, coach, counselor, and friend.

People in the firm liked him, saw that he had talent, but never knew what to do with him. He more or less wandered around the firm but never felt that he was part of it. He did not know enough about the consulting world to make his needs felt clearly; the firm did not understand him or what they wanted to do in Asia.

He finally left, joined another management consulting firm in the States, experienced the same kind of alienation as before, quit, and joined an even larger consulting firm. This time he was sent to China where he became a senior consultant in the firm's "human capability" practice.

This job, too, had its ups and downs. The office was headed by an American who did not seem to understand the business and management needs of China. He was at the end of his career, so he wanted peace and quiet and stayed aloof. China was the wrong place for that. As part of a marketing effort, Cheung gave conferences on human resource issues around the country. People noticed him. Soon he was offered the job of starting and running a government-funded human resource institute in Shanghai. He took the position and finally found a place where his creative juices flowed freely. Now that he has a few years of managing a company under his belt, he is thinking of giving up that post and becoming a solo entrepreneurial consultant to senior managers. Cheung knew in a broad way what he wanted in terms of a career but the clarification of his career goals and his path toward those goals involved a great deal of experimentation. He is still on his journey. This case highlights an important fact. Planning need not take place all at once. It is often enough for clients to develop a general plan which they then translate, with or without the help of a counselor, into a series of "experiments" focused on the goal or goals they have drawn up for themselves.

COMMUNICATION SKILLS

I address communication skills last for a reason. Some people mistakenly see the communication skills outlined in the *Skilled Helper* as the helping process itself. Communication skills are essential for building the helping partnership and for helping clients move through the stages and tasks of the helping model. But they are essential tools for making the model work and not the model itself.

If the helping relationship is a partnership, then the interaction between client and helper needs to be a *dialogue*—a solution-focused, helping, and therapeutic dialogue. Helpers need the communication skills that contribute to effective dialogue for two reasons. First, they need the skills to actively engage their clients in the give-and-take of the therapeutic dialogue. Second, since many clients are not good at dialogue, helpers need these communication skills to help clients whose communication skills are weak to participate actively in the dialogue. Effective dialogue has four characteristics:

1. **Turn taking**. Effective communicators do not engage in monologues. Nor do they sit idly by while their partners engage in monologues. I speak, you speak.

2. **Connecting**. Effective communicators avoid intersecting monologues. They connect. What I say responds to what you have said. And I expect you to do the same. Effective dialogue is organic in nature.

3. **Mutual influencing**. The participants in effective dialogue influence each other. I am open to being influenced by what you say to me and I expect that you will return the favor.

4. **Co-creating outcomes**. By engaging in steps 1, 2, and 3, effective communicators end up co-creating the outcomes of the conversation. As a counselor, I don't tell my clients what to do, but I do provide guidance for their journey.

The basic communication skills needed to engage in effective dialogue are discussed and illustrated extensively in the *Skilled Helper*. They include such skills as:

- listening actively and empathically to clients in the context of their lives.
- actively responding to what clients say with understanding.
- conveying information and other messages clearly.
- asking questions and using other probes to help clients explore issues.
- helping clients identify and explore both cognitive and behavioral blind spots.
- challenging clients to see alternatives.
- challenging clients to look at the consequences of their behavior.
- using these skills to help clients become constructive partners in the dialogue.

Interpersonal communication is culturally conditioned. Rules abound and they differ from culture to culture. Consider the following example.

I once gave a conference to a Catholic bishop and some two hundred seminarians in the middle of Tanzania. I began by saying something like this: "I'm going to teach you a model of helping, but you are going to have to tell what works and what does not work here in Tanzania." And off we went. In the debriefing some days later I learned that they had few problems with the model itself, but they had some problems with the communication skills, at least as presented. The bishop himself gave me the following example.

Two men, neighbors, were having a dispute about their respective properties. One day, one of them, frustrated by the lack of response of the other, barged into his house and began confronting him with a variety of complaints. Since the other seemed almost indifferent to what he was saying, he finally yelled, "Look now, you're not even listening to me!" The other man looked up and said, "You have not greeted me." The man who had barged in, stunned, said, "Why, of course." They greeted each other and then got down to business.

Here is an example from one of my classes. I once asked a student of mine from Sri Lanka to be my counselor. I wanted to show the class how a client would move forward in the helping process only to the degree that he or she was actually being helped. At the end of a ten-minute session we stopped and the student received feedback from the other members of the group. One feedback theme was that he had not kept very good eye contact with me. When it was his turn to respond to the feedback he said, "In my culture we do not look a person in authority in the eye." We all learned something about the culture of communication. While keeping good eye contact is in Western cultures is usually a sign of respect and engagement, it is wrong to assume that this is universal.

All communication skills—active listening, responding with empathy, asking questions and using other forms of probing, challenging, explaining, informing, giving advice—are conditioned by both the cultures in which they take place and by individual differences within the culture. Once I confronted an Arab friend of mine who I thought was yelling at his mother on the phone. He said, "I wasn't yelling at her. I was yelling at the issue. That's the way we do it." I also came to realize that his readiness to give advice to almost anyone on any issue was not the kind of advice giving that I personally try to refrain from in counseling. He said, "In my culture we are always getting advice from everyone. It's their way of saying that they want to help you. In the end, we pick and choose what seems best. No one is surprised when you don't follow their advice. They are just telling you what they think."

Rich McGourty works with a group of physicists. He relayed this incident to me.

> The conversational norms which govern scientific debate among particle physicists can be quite surprising to an outsider. I overheard one say to another in the middle of a discussion, "This makes no sense!" To my mind, that would have signaled the end of civil conversation. But it didn't and later the objecting physicist reassured me, saying, "Oh, that. No, he didn't mind. Here saying you think something's wrong or even that you think someone's a bit crazy just means, 'Let's talk.'"

These scientists have their own subculture. So if you're counseling a particle physicist, you know what to expect.

I have learned a great deal about communicating in a cross-cultural situation from my mistakes in various settings. I worked as a consultant with the managers and professionals of a number of different international agricultural research centers scattered around the world. The work included a great deal of coaching and counseling. I remember a meeting at one of the centers vividly. A newly-arrived American researcher spoke about how he thought that local workers were not being treated right. This was his first job abroad. He spoke angrily about the "injustice" he saw around him and complained that "no one was doing anything about it." The other members of the group said nothing. I got angry seeing one of my fellow Americans arrogantly and patronizingly speaking for the "downtrodden" staff and field workers without knowing what the culture of the country or the culture of this institution was really like. Frustrated by his arrogance, I took him on, thinking that I was doing the group members and the institution a favor.

Later on, however, I found out that the members of the group—they came from many different countries around the world—were shocked and embarrassed by my intervention. Such public "reprimands" were not allowed in some of their cultures. But, even more to the point, publicly challenging one of the professionals in this particular international research center was considered very poor form. I was the cultural culprit. Or both of us Americans were. Had I to do it over again, I would have taken him aside after the meeting and engaged him in a discussion of both the issues and his angry public confrontation of the leadership of the institution. The group members knew that he was talking out of turn and that he needed to be set straight. But I was not the person to do so and a training conference was not the proper setting. Challenging or confrontation is strong medicine. The cultural and personal context needs to be factored in right from the start.

I received an unexpected bit of reinforcement regarding the usefulness of the kinds of communication skills discussed in the *Skilled Helper*. I was in Beijing talking with a couple of people from the Ministry of Education about the need they saw for counseling programs in schools. At the beginning of the conversation they asked me about my experiences in China. Among other things I told them about how many people I met that wanted to get on the new-economy bandwagon and the strategies they were using to do so. For instance, I had given a conference to a group of business-school professors in Shanghai. I had shared with them some of the models that I use in consulting with organizations. After the meeting one of the professors put his arm around my shoulder and started talking about the kind of consulting business he and I could set up in China. The officials from the ministry listened to what I was saying with great attention. Then, through the interpreter, they asked, "How have you gotten to know us so well so quickly?" I was so startled by their question that I started giving disclaimers. I was not sure that I had really gotten to know my Chinese colleagues that well or that quickly. But then I paused and said that good counselors are always learners. I named some of the communication skills discussed in the *Skilled Helper* and said that these were my tools. I also mentioned the "shadow-side" model (Egan, 1994) I use to "get behind the scenes," as it were, in order to understand what is really going on in both personal and organizational behavior. Shadow-side behavior is universal but it has many different cultural expressions.

And so ends our brief journey through the world using the *Skilled Helper* as our lens. Or was it a wander through the *Skilled Helper* using the world as our lens? I have tried to demonstrate how this helping model together with the methods, communication skills, and values that make it work have made me a good multicultural counselor, probably long before the multicultural movement got up its current head of steam. The *Skilled Helper*, though it lies in the rich cognitive-behavioral area of therapeutic psychology, owes much to the person-centered approach pioneered by Carl Rogers (Kirschenbaum & Jourdan, 2005). It is about basics that can and need to be used with any approach to helping. Because it is a problem management and opportunity development approach to helping it has universal appeal. Everyone is the world faces problem situations; everyone in the world needs to identify and develop opportunities in order to grow as a human being. That's why problem solving is a human universal and enjoys such an enormous research base. Because of its universality it has been easy, in my experience, to adapt it to cultures around the world. In the end, I know that you are going to use your understanding of the *Skilled Helper*, diversity, and multicultural competence in your own way. But your success will always lie in the improvements that your clients make in their lives.

REFERENCES

American Psychological Association. (2003). Guidelines on multicultural education, training, research, practice, and organizational change for psychologists. *American Psychologist, 58*, 377–402.

Bronfenbrenner, U. (1977). Toward an experimental ecology of human development. *American Psychologist, 32,* 513–531.

Chen, S, & Davenport, D. (2005, Spring). Cognitive-behavioral therapy with Chinese-American clients: Cautions and modifications. *Psychotherapy: Theory, Research, Practice, Training., 42*, 101–110.

Darcy, M., Lee, D., Tracey, T .J. G. (2004). Complementary approaches to individual differences using paired comparisons and multidimensional scaling: Applications to multicultural counseling competence. *Journal of Counseling Psychology, 51*, 139–150.

Day, S. (2005, May). Innovations in counseling. *Counseling Today*, 37.

Diamond, J. (1997). *Guns, germs, and steel: The fates of human societies.* New York: Norton

Economist (2004). The shock of the comparative. January 17, p. 36.

Hansen, N.D., Pepitone-Arreola-Rockwell, F., & Greene, A. F. (2000). Multicultural competence: Criteria and case examples. *Professional Psychology: Research and Practice, 31*, 652–660

Kirschenbaum, H. & Jourdan, A. (2005, Spring). The current status of Carl Rogers and the person-centered approach. *Psychotherapy: Theory, Research, Practice, Training., 42*, 37–51.

Kia i Kitaoka, S. (2005, January). Multicultural competencies: Lessons from assessment. *Journal of Multicultural Counseling and Development, 33*, 37–47.

La Roche, M. J., & Maxie, A. (2003). Ten considerations in addressing cultural differences in psychotherapy. *Professional Psychology: Research and Practice, 34*, 180–186.

Moffett, M., & Samor, G. (2005, May 24). In Brazil, thicket of red tape spoils recipe for growth. *Wall Street Journal*, A1, A9.

Murray, H. A, & Kluckhohn, C. (1953). Outline of a conception of personality. In C. Kluckhohn, H. A. Murray, and D. Schneider (Eds.), *Personality in nature, society, and culture*. 2nd Ed. New York: Knopf.

Pope-Davis, D.B., Coleman, H. L .K., Liu, W. M., Toporek, R. L. (2004) *Handbook of Multicultural Competencies in Counseling and Psychology*. Thousand Oaks, CA: Sage.

Stuart, R. B. (2004). Twelve practical suggestions for achieving multicultural competence. *Professional Psychology: Research and Practice*, *35*, 3–9.

APPENDIX

The Tilford Group
WORKING MODEL FOR FURTHER RESEARCH ON MULTICULTURAL COMPETENCIES

I. KNOWLEDGE

- **Cultural Self**—The ability to understand one's ethnic identity and how it influences identity development.

- **Diverse Ethnic Groups**—Knowledge of diverse ethnic groups and their cultures.

- **Social/Political Frameworks**—Awareness of how economic, social, and political issues impact race and ethnic relations.

- **Changing Demographics**—Understanding population dynamics related to ethnic minority and majority citizens.

II. PERSONAL ATTRIBUTES—Traits needed by those who live and work in a diverse world.

- **Flexibility**—The ability to respond and adapt to new and changing situations.

- **Respect**—An appreciation for those who are different from one's self.

- **Empathy**—The ability to understand another person's culture by listening to and understanding their perspective.

III. SKILLS—Behaviors and performance tasks needed to live and work in a diverse world.

- **Cross-cultural communication**—Verbal and nonverbal communication skills in interaction with those who are culturally different from one's self.

- **Teamwork**—The ability to work in culturally diverse groups toward a common goal.

- **Listening** —The ability to attend to what others are saying.

- **Conflict Resolution**—The ability to resolve cultural conflicts that occur between individuals and groups.

- **Critical Thinking**—The ability to use inductive and deductive reasoning.

- **Language Development**—The ability to speak and write more than one language.

- **Leadership Development**—The ability to provide multicultural leadership.

PERSONALIZING EXERCISES

1. Describe culture. What is the "thinking" part of culture? What is the "should" part of culture? What is the "doing" part of culture? What is the importance of the term "shared" in the description of culture? What is the importance of the term "patterns"?

2. What are some of the "thinking" dimensions of your family culture? What are some of the "musts" in your family culture? What are some of the "doing" parts of your family culture?

3. Consider a sibling or another relative or close friend about your own age. Name a half-dozen ways in which you are similar. Name a half-dozen ways in which you are different.

4. Name some ways in which your differences enrich your relationship. Name two or more differences that at times cause difficulties in your relationship.

5. How are the terms "diversity" and "culture" related?

6. What is meant by the term "sense of the world"?

7. How would you describe your current sense of the world?

8. Think of someone you think has developed a sense of the world. What is this person's sense of the world like? How does it affect his or her thinking and behavior?

9. How does the social justice dimension of multiculturalism relate to people's sense of the world?

10. What would be different if your community—you yourself, your family, friends, classmates, colleagues at work, and so forth—had a highly developed sense of the world?

11. After reading the brief description of Chester, how would you feel if you found out that he was going to be “assigned” to you as a study partner or your partner at work? What would be the advantages? What misgivings might you have?

12. Think of someone—yourself or someone else—who has a problem or some unused opportunity in which ethnic culture plays an important role. Describe the part culture plays.

13. Describe some of the things you would have done had you been Maria’s counselor. What do you think some of your challenges would be?

14. Describe your understanding of the term “personal culture.”

15. What are some of the key dimensions of your own personal culture? Include dimensions of ethnic and family culture if they are key.

16. Describe someone who is close to you in terms of his or her personal culture. In what important ways do you and this other person differ in terms of personal culture?

17. Consider someone you know fairly well who has some kind of problem in living. If possible, choose a problem that involves a relationship with some other person. Describe that person's problem situation in terms of relevant dimensions of his or her personal culture. In what ways, if any, are dimensions of ethnic culture involved?

18. What do we mean when we say, "We counsel individuals"?

19. Read the list of cross-cultural competencies outlined above. Which of these would be easiest for you? Name a couple of them and explain the reasons behind your choices.

20. Which of these would more difficult for you? Name a couple of them and explain the reasons behind your choices.

21. Why does Stuart say "it is never safe to infer a person's cultural orientation from knowledge of any group to which he or she is believed to belong"?

22. When you meet a person for the first time, in what ways do you describe yourself as a "learner"? What does humility have to do with it?

23. The Tilford Group operates on the principle that multicultural competencies are for everyone. Why do they say that? How do you think other forms of diversity fit into this kind of thinking?

24. How would you world change if all the key people with whom you interact had practiced the multicultural and diversity competencies outlined in the brochure and in the Tilford Group Appendix? In what ways would things be better? In what ways would things be more challenging?

25. If problem management and opportunity development are tasks for people around the world, why should a problem-management and opportunity-development model be itself scrutinized through a multicultural lens?

26. Why do you think I was surprised to find strong women in positions of authority in Muslim universities in Malaysia? What does this say about stereotypical thinking?

27. Comment on my failure to follow the advice of my Chinese translator when I was giving a conference in Shanghai. When is it allowed to challenge a culture?

28. Comment on this statement: "Cultures can be breeding grounds for blind spots." Give an example of a blind spot found in a family culture—either in your own family or one you know well.

29. Why are communication processes so culturally sensitive?

30. What issues are difficult to talk about either in your own family or in a family you know well? What is it about the family culture that makes it so difficult to talk about these issues?

31. What role does dialogue together with the skills that enable it play in multicultural competence?

32. Review the four dimensions or principles of dialogue. In what ways might each be culturally sensitive? What cultural assumptions are embedded in this description of dialogue? What might some form of authentic dialogue look like in a highly authoritarian culture?

33. What incentives do you have to become a diversity and multi-culturally sensitive and competent person?